I0476530

VIDEO MARKETING

How To Produce Viral Films And Leverage
Facebook, YouTube, Instagram And
Twitter To Build A Massive Audience

Entrepreneur Publishing

COPYRIGHT

Disclaimer

FREE GIFT
Kindle 5 Star Books

Free Kindle 5 Star Book Club Membership

Join Other Kindle 5 Star Members Who Are Getting Private Access To Weekly Free Kindle Book Promotions

Get free Kindle books

Stay connected:

Join our Facebook group

Follow Kindle 5 Star on Twitter

Also, if you want to receive updates on Entrepreneur Publishing's new books, free promotions and Kindle countdown deals sign up to their New Release Mailing List.

For entrepreneurs: http://www.entrepreneurfinesse.com

Table of Contents

Introduction

Step Number One: Release Your Focus On Quality

Step Number Two: Choose And Target An Audience

Step Number Three: Keep The Subject Simple

Step Number Four: Give The Subject Center Stage

Step Number Five: Include Your Call To Action

Step Number Six: Get The Video Out There!

Step Number Seven: Create Contemporary Content

Step Number Eight: Create. Promote. Release. Repeat.

Conclusion: Do's and Don'ts

Specific Details & Common Terms In Video Production

Introduction

It seems these days, everyone is familiar with the concept of a 'Viral Video'. The common thinking is that if you are at the right place, at the right time with a cell phone camera or Video camera, all you need to is capture the images, slap them onto YouTube, and if you are lucky, you get a video out there that goes viral, and suddenly you are a millionaire.

Let me ask you something. Do you want to make a lot of money? In particular, money that is generated while you are busy doing other things, that does not depend on your particular and personal efforts to produce? To answer for you, in all candor, if we are honest about it, we all do. There is something very appealing about an income stream that does not depend on active work, because we already do that. We know that work equals pay, in most cases, so pay that comes from other resources is highly appealing.

In this book, we will look at the means by which one can create rising income, through the creation of and access to desirable and potentially viral videos. By understanding the value and the power of quality content in digestible image duration, we create a passive income that will go on and on, and could become quite profitable.

We will go over the kinds of videos draw attention; how to position and develop a marketing campaign that will put eyeballs on your material, and the ways to make the information really compelling.

We will highlight eight different facets of this business that will provide key information that can open the purse strings, and generate increasing and stable income to you, if you follow them all and make the efforts necessary. In the end, you will have a solid lock on the critical tools that can create real value of video marketing, as well as a great vocabulary to begin your foray into this vibrant and exceptionally alluring environment.

So jump on it, have a good read. Bring along camera and some cats (you'll see why shortly) and let's build you compelling viral video library, shall we?

Step Number One: Release Your Focus On Quality.

In this chapter, you will learn:

- Getting the best result doesn't need costly investment.
- For videos, the type and purpose define the quality of performance.

Quality video doesn't mean costly equipment.

If you have spent any time at all watching those catchy, interesting, clever 'viral videos', you have seen a wide spectrum of quality, lighting, performance, audio clarity, and other qualifying criteria. Some of the videos are grainy and confusing, others are studio-quality masterworks.

In all frankness, there is no particular level of appropriate quality that ends up as a minimum for the development of a positively catchy and appealing video clip.

Clear images don't require expensive cameras or lighting.

For this reason, if you have a fairly modern cell phone with a camera, you are probably already set to shoot video. The appeal of both snapshots and short video, for personal, family, and community purposes have led the various producers into a product war of sorts, ensuring just about any camera out there would be sufficient for most tasks. As we proceed, certain situations will present themselves where this may not be the case, but we will deal with those as we find them in the coming chapters.

Sound, too, is usually more than adequate.

Many of the kinds of videos that 'go viral' either have little or no audio characteristics, but even those that do, because again the primary photo and audio device is a phone, the particular sound capture requirements are already adequate for general use. We will cover other particulars in later chapters on this account as well.

Keep it Concise

"Brevity is the soul of wit." Most of us know almost instinctively that the best of the 'viral videos' out there are short and sweet. Some basic ideas about the very length of the video you might want to create do entail keeping it short. Here are some tips about how to do just that.

Cut to the Chase

When scripting the video, if it is operating as a script, cut out all the lead-in and trailing video, but the scenes you want to remain. Replace long explanatory video with a scrolling text explanation, and run that just before the particular scene. When the video is complete, cut in a scrolling text with your call to action and any explanation you think it needs, as well as concluding with a "Thanks for watching" message.

Subtitles go a long way.

Often, it is easier to just run a small scrolling text along the top or the bottom of the image, as a means of explaining things to the viewer as that seems to help the continuity flow. If necessary, include some humor or irony in the subtext, to cause them to watch the video more than once.

Realize the breadth of your options

At this point, be excited about the potential, and as you read the following sections, your alternatives and options will only improve even more than you may have considered. Remember, the field is really relatively new, and there are a lot of different choices about not only your video subject, but also your video theme, method, angle, etc. Get ready to have a lot of fun!

Keep a Positive, Uplifting, Inspiring theme

Fundamentally, the aspect of videos that drives the almost manic viewing is the ways in which the images and the stories behind them are matters of positive attitude, good stories, and happy endings. Even the stories or images that have a dark or challenging theme tend to be bright, happy imagery.

We all want to win.

By keeping the focus on the positive, you continue the very flow of positivity that brought the viewers to your video. Play that optimism and enthusiasm to its core, and keep the viewer engaged.

One good video leads to another.

When the viewer enjoys one of your videos, it incites them to try another. And another. Keeping that momentum going leads to multiple views, increase engagement, and returning viewers. All lead to more page views, and that can, over time, lead to payoff through the variety of monetizing methods, which we will discuss later.

Light-Hearted and Easy to Identify with Experience

Think of your own video-watching. Are you more likely to watch one about a cute animal, or a how-to on flossing your teeth? Sure, we might try to suggest that we think about our dental condition more, but when push comes to shove, the cute sneezing panda or caterwauling panther cub is going to eat up that spare minute. So create your video with an eye to what the viewer desires, and you will be ahead of the game.

Eyeballs equal sales

There are two things to remember when creating your video content. The eyes of your target customer. Though we haven't touched on it yet, it is imperative that the attention you draw is turned into a financial benefit. You don't make a video for the sake of having video. You do so to drive the consumer to act, if it is nothing more than watching another video. Because attention is so precious, you never want to give it away without some action you want to be completed.

The viewer is a buyer

It may not feel like it, particularly if your message is a 'soft target' – getting the consumer, the viewer to press the 'like' button, or to 'share' the video through social media. Still, whether they are writing the check for the end product, or opening their intellectual wallets by conveying your message on into their networks, the viewer is a customer, and we do well when we remember that.

Step Number Two: Choose And Target An Audience

In this chapter, you will learn:

- How to identify and target your marketplace effectively
- Establish a basis for not just one video, but a whole library.

Choose your Trigger or Triggers

Depending on whom you are reaching out to, the kind of things that interest or appeal to them will vary considerably. When deciding what kind of video you want to create, you do want to take into account precisely whom you are hoping to attract. For instance, if you are promoting a clarifying ointment for skin wrinkles, it would be doing your product a disservice to draw viewers from a teen or early-twenties audience, or if you are promoting a hip-hop album, you wouldn't necessarily want a large portion of your audience to be middle-aged cattle ranchers. Even so, here is a short list of possible video triggers, and potential audiences they may attract. Of course, your mileage may vary; there are some wacky videos that defy the normal rules, of course.

Random (non sequitur)

These videos tend to do something extraordinary, but that are not able to be clearly associated with other topics. Great for younger audiences, or for general audiences that usually are so niche as to be untenable as a demographic on its own.

Surprising

The 'Jump out of your skin" kind of videos, great for young women who seem to really get a charge out of the effect, and people who study such, because it gives them new grist for the mill. Usually also good for one-offs. If all your videos follow this mode, you won't have a very deep following.

Sexy

Best for the singles crowd, and generally should be created in tandem; some are drawn by female lead, some by male, so making duplicates offer you the chance to get both marketplaces. Certainly more optimal for products that are not targeted to families or youngsters.

Comical,

Although laughs would seem like a logical form for just about every product and situation, there are a great many variables, from the focus of the humor, to the kind and type to be uses, as well as the resolution and ambiance of the result. Use with caution. And a rubber chicken.

Shocking,

Similar to the surprising category above, keeping your video within the narrow band of tastefulness and visceral response is a big and wide-ranging conditional mine field. Many of these are special-effects or otherwise manipulated video. Use with a mind to restriction, so as to not sully your site's credibility overall. As my grandfather would say, make the effect so outrageous no one will believe it... that way no one will have reason to question your overall credibility.

Spectacular

One of the particular areas wherein the video quality can, and will, play a big part of the impact, A spectacular video, with all the bells and whistles meant to draw the viewer into the story, will certainly be better received if it is made with a higher quality presentation than otherwise.

Establishing a Theme.

When you are beginning the video aspect of your operation, you can try a lot of different things, different themes, and different types of videos. Not only is that natural and acceptable behavior, it gives you a wider balance base for your ongoing video creation. Nonetheless, over time, you will develop or discover that you have a particular penchant or talent for a particular type or category of videos. Doing so, you will even find your video catalog becoming more useful over time to bring in and keep your audience occupied.

Here are some Types of Video Themes you may want to consider.

- o EXPLAINER/TRAINER VIDEO
  Your product or service, explained for the uninitiated in a humorous or enthusiastic way.

o WEBINAR - SEMINAR
Briefings and information in bite-size pieces for everyone to learn a little

o PROJECT OR PRODUCT REVIEWS
Identify and associate your goods or services in the realm of the consumer. Generally positive and upbeat, not critical.

o VIDEO PSAS OR ANALYST REPORTS
Support your overall industry categorically, for industry identity and a positive external company and product image.

o CORPORATE CULTURE OR INVITATION VIDEO
Bring new clients in, and make ongoing customers feel at home with your company and product.

o TESTIMONIALS OR FBI'S
Allow your good customers and new clients to help you reach out to still more clients and potential customers through their experiences

o ANIMATIONS
Particularly if the images and humor support your brand, your corporate identity, and your collective corporate image, these can become iconic.

o MAN ON THE STREET INTERVIEWS
Used in association with testimonials and informational pieces, these can amplify the effectiveness of your other efforts if done professionally, and without self-deprecating or industry-effacing humor or deportment.

o CHALLENGE VIDEOS
In keeping with your overall message, these call-to-action videos can amplify your effectiveness and brand your operation as a solid support for your industry.

o PARODIES OR PARALLELS
Improve your visibility by taking advantage of what is working for others. If you want particularly effective parody, trigger these on your own successes, to create a montage of the success for others to emulate.

○ MOCKUMENTARIES

These are a great way to get your message out with humor and wit. With an eye toward the duality of actual information with a humorous bent, these can be particularly effective in conveying large chunks of normally snooze-worthy information.

○ STATE OF THE MARKETPLACE UPDATES

The purpose of a message like this is to get out the details for your supporters, and to create enthusiasm for your business. If the data is particularly bleak, consider the Mockumentary angle, to put a more enjoyable feel to the data, rather than simply reporting the facts.

Step Number Three: Keep The Subject Simple

In this chapter, you will learn:

- How to 'boil down' your message to the minimum size.

- How to drive people to view even your more educational videos.

Make It Short.

- o REDUCE SCENES TO THE MINIMUM NECESSARY
  While this may seem difficult to calculate, it falls squarely into that category of "you'll.

- o REDUCE CAST TO LIMIT 'TAKES'
  Checking the real estate values, number of houses listed for sale, and other factors can

- o SHOOT IN AS FEW 'TAKES' AS POSSIBLE
  Census values can help in determining the stability and income levels in the Low to

- o USE "TEST SHOWINGS" TO ENSURE YOUR MESSAGE IS GETTING ACROSS.
  While this may seem difficult to calculate, it falls squarely into that category of "you'll.

- o USE SPEED CREDITS, IF CREDITS ARE NEEDED
  Checking the real estate values, number of houses listed for sale, and other factors can

- o CONSIDER SERIALIZING VIDEO
  If the resultant video is still over-long, consider serializing the shots, to make multiple videos in a series. Doing so, if properly annotated and linked, can make a longer-form video more palatable, and easier to digest by viewers.

Make it different or unique

There is no set means by which to differentiate one topic from another, or to guarantee what you create your video will automatically appeal to any given demographic, let alone be the specific one you were targeting. Instead, consider these tips to make your video far more likely to be appreciated, than disdained.

Ways to make your video content compelling.

○ LOOK AT THE TOPIC IN A FRESH WAY

IF the concept of your video sounds boring to you, or to your test audience, it probably will be. Spitball some wacky or at least off-kilter alternatives, and when you come across the right approach, you and the would-be audience will both feel it. Humor is not hard, if you don't try to force it. Lightness of the topic helps, so consider increasing the levity of your premise if nothing else has started working.

○ SHOW THE TOPIC FROM A DIFFERENT POINT OF VIEW

You know your topic from the side of the writer. Consider the same from the standpoint of the consumer. The actor in the video. The bystander. Maybe even the microphone or the camera. The point is, often what works to convey a message is something beyond the obvious.

○ TAKE CERTAIN ASSUMPTIONS TO THE LOGICAL EXTREME

What if the stamp on the letter had to get bigger when it got more expensive? What if inflation of tires had to match financial inflation? Any pertinent, but ridiculous comparison can create fun, interesting, and even poignant conditions that will help to drive your video's point home.

○ REPLACE ONE OR MORE PLAYERS WITH AN ANIMAL OR INANIMATE OBJECT

If the guy in front of you in traffic really was a turtle, what would that do for your day? Sometimes, the other guy in the line in front of you really IS a rock. By creating clever video conditions like this, you can make a challenging subject a little easier to understand, or a rougher situation more palatable. Just be open to the options.

○ ONE WORD: ANACHRONISM

If you were selling your concept to a Balinese dancer, what would that do for your sales? What if it were a Russian Cossack, or Attila the Hun. IN all of these cases, it is the situation that makes for humor or at least increased interest in the information.

Give it a Gripping and Interest-Driving Title

We haven't shot a single scene of the video, but we need to begin to put together the concept for presenting and promoting the clip well in advance. By choosing keyword-driven titles, or current-affairs drivers in the first place, we can build and maintain a longevity to the video, beyond just its dry and subject-driven basis.

Tips to Create the Perfect Title

- ASK OR ANSWER A QUESTION

  People are puzzle-solvers and solutions-hunters. If you create a question in the viewer's mind, and then either answer it or challenge them to, you will get them to follow the link, if to do nothing but validate their beliefs. The greater the challenge, sometimes, the greater the response. The title, in essence, becomes a call to action.

- DON'T BE AFRAID OF SUBTITLES

  Whether you extend the actual title to include it, or just use the video description space, the keywords are still a function of the search for the video.

- USE CORE KEYWORDS WHEN POSSIBLE

  In cases where the links are not attractive, consider bitly.com to make the title one that can include your keyword, even when the title of the video does not.

- DELIVER ON YOUR TITLE

  Whatever your content, make sure the video delivers what you say in the title, because your credibility will impact how many of your videos a given viewer will look at. If you fail to deliver, you shoot yourself in the foot for future projects with the client who came to see what your title promoted.

- BE CONTROVERSIAL OR CHALLENGING

  The great thing about controversy is that the viewer will appreciate either side of the issue, so long as again thee title matches the video content.

- CHECK ON ORIGINALITY

  There is nothing wrong with having a title someone else shares, except it makes your video that much harder to locate and differentiate. So do a little due diligence, and look for a unique way to say the same thing, if necessary. Being original can be a good thing, especially if your content helps to define the particular issue overall.

Step Number Four: Give The Subject Center Stage

In this chapter, you will learn:

- What elements you will want in your video, regardless of type or topic
- What kinds of video categories are the most helpful

Experience Videos

People watch videos for a million different reasons. Some are sheer escapism, others a spot of joy, and still others have deeper, more lasting meaning. When you begin to create your videos, be aware of what the real message is you want to send, and ensure that what you present has the impact you desire. If there one truth about presenting information through visual means, the interpretation of the viewer needs to match what you intend, or else your result will not be the response you were looking for.

Talents and Skills

A primary purpose for a video is to present the evidence that you can accomplish what you intend. Using this video message to demonstrate the talent, or to draw attention to the value or skill or unique method all have an intrinsic teaching value. Adding entertainment, cleverness, or snappy music can change it from just another snippet of video, and turn it into sensation-worthy viewing.

Running Commentary

Very popular 'as you play it' videos suggest that viewers might get a kick out of video demonstrating the everyday in a more energetic way. Consider using some kind of commentary/monologue to make the routine seem to be so much more.

"How To" Videos

Not everyone has the skill sets and talents you have. By creating videos that share your secrets, your special knack for something that they, too, could do if they chose to, will create a fascinating montage of instructional and informational videos.

Specific and desired skills

There are particularly interesting videos to be made, if the job or task you want to train is particularly visual, or if you have a unique or fascinating way to accomplish it. Even basic tasks can be made entertaining, if you add humor or drama to the video.

"What's Behind the Curtain"

Another aspect of video that can be done, is when the actual processes for something that is less well known is brought to the viewer in a way that shares secrets, surprises, or supplemental information not normally available. The concept of an upcoming revelation is sure to pique interest, particularly if there is a coherent buildup of excitement before the actual revelation is made.

Blasts From The Past.

As the Internet generation comes of age, there are a great many things about the past that are simply not common knowledge, By bringing these tidbits of information to awareness, the new generation learns more about its past and heritage. Whether it be old hardware that no longer is needed, old ways of doing things that have fallen out of favor, or even old terms and speech patterns that are no longer in vogue, such make great, informative video fodder.

Super Spontaneous Shots

Beyond the well-choreographed videos we have been speaking of, there are the more endearing, seemingly spontaneous shots, that bring our humanity nearer, make our appreciation for life itself dearer. Here are some tips, to help facilitate capturing these gems that can lead to more appreciation and viewership.

○ KEEP YOUR CAMERA HANDY

While this is perfectly natural for photographers, it isn't always for anyone else. Still, the Cellphone cameras of today are downright amazing in terms of imagery, and even the sound collection on many of them is more than sufficient for YouTube or other video outlets.

○ USE THE LONG ZOOM.

Capturing candid imagery is all about the targets being unaware the cameras are rolling. By shooting at a reasonable range, you keep the shots natural and the action central to the filming space.

○ KILL THE FLASH

Shooting action footage is particularly good if it seems no one else is there with a camera, that yours is the one and only true shot. So try to avoid making video that shows other cameras in the shot.

○ TAKE A LOT OF VIDEO

Sometimes it is the video taken between the formal videos that is the most entertaining and surprising. So keep the camera rolling. One can always cut the boring and the bland, but if you miss the special moment because the camera was off, it is awfully hard to recapture precisely as preciously again.

○ POSITION YOURSELF STRATEGICALLY

Taking shots in the usual way will result in usual videos, so consider hanging the camera from a tree or scaffold, or running by the action on foot. The way in which you place and frame the shot will have a lot to do with its uniqueness and appeal.

○ TAKE SHOTS 'WHERE THE ACTION IS"

Shooting something amazing from across the room is cool, but for really engaging footage, consider the camera as the middle of the action, and taking the camera angles and shot framing into the center of the fray, to get a surround-sound kind of feel, like the action is on every side. People do this for concert footage, but imagine if you had it at a business event or public event.

○ CAPTURE THE CONFRONTATION/CONFLAGRATION

While this may seem difficult to accomplish, think of the daredevil that mounts a camera on his head for a jump, or on the frame of his bike for a different shot. In like fashion, you can make even a family video more appealing if it is taken from the middle of a group hug, or from the chapel above the wedding. Just consider all the possible ways to capture the experiences, and go for the ones that are the most innovative, the most edgy.

○ SHOOT FROM ASSORTMENTS OF ANGLES.

Remember the old 'Batman" shots, taken from an angle to give a different frame of reference? So, too, consider the various ways your shots can be made, and take the time to experiment in advance as well.

o SHOOT FROM UNEXPECTED PERSPECTIVES

There's nothing more refreshing than seeking a traditional video shot from an interesting and refreshing point of view. Maybe as an ant on the front row. Or suspended behind the artist, to capture their moves from the other side of the stage. Thinking outside the box, beyond the sphere, is what will make your video stand out from the competition.

o CAPTURE THE OUT-TAKES

Keep the camera rolling, and get more shots on the center of the action than you will ever need. The fun stuff is found in the moments of humanity, when all the orchestration and choreography fails, and people are just people. Those are the moments that are most precious, most appealing.

o RE-FRAME

Think also about the direction of the video's story Think of what the rewind might look like, or the super slow mo. In all these cases, there are minutes of pure video gold to be investigated.

o CONSIDER THE STORY IN THE PICTURES

Adding to the comment above, by thinking in terms of composition, you can make a story into a masterpiece by just taking into account the drama that the action can represent, and taking it to the next level as well.

Step Number Five: Include Your Call To Action

In this chapter, you will learn:

- The key to the Action Phase of your video plan
- The power of post-production

Taking Advantage of the 'Bump'

Regardless of your   strategy, whether planned video or an attempt at an attention-getter, you want your video to have a purpose – a call to action. Maybe you are introducing a product. You need folks to visit a website. You want an email sent to a Congressman.

All these different ideas are still calls to action. What we are talking about, then, is how to actually enact such a triggered effect from the viewers. In a project guide like this, there are far too many variables, so we can't possibly cover them all. Nonetheless, here is a fairly comprehensive list of Call to Action possibilities, divided into categories for your consideration.

Self-driving Calls to Action

Lead Generation

Generally, the viewer agrees to offer you their email address, with the understanding you can contact them thereafter.

Forms Submission

More specific than lead generation, the forms submission call to action usually leads to a contract allowing you the particular information you request of the viewer. Can be extensive, depending on the form.

Learn More

A click through grants the viewer additional information. Can lead to a product info page, service guidelines page, or other resource.

Testimonial - Gathering

Retrieving responses from those who have purchased the product, used the services...

Process Feedback

Requesting feedback from those going through the process, to build on former success in order to garner new success.

Catalog or Service Directory
More extensive and practical than just information, usually this links directly to sales functions of the site associated with the video link.

Social Sharing
Considered basic, this generally has a multi-tiered option, from just identifying with the image, to sharing the image into a social media group, or even embedding the video and link into the consumer's own communications network.

Customer Retention
With a sign-in feature, your current clients can benefit additionally by following this kind of link, that at once identifies them with the particular video, and rewards them ins some fashion for their effort and viewership

Event Promotion
This kind of CTA brings the viewer to information and perhaps an RSVP function, which can afford you good intelligence data on which videos bring them to the event and the like.

Sales Close
Everybody's favorite, the final button that guides the consumer to pull the trigger, make the purchase, order the service, etc.

Post-Production: Where the Rubber Chicken Meets the Road
Now that you have a clear idea of what you want your viewer to do when he is finished watching your video, how can you make that easy for him.

The calls to action still need the viewer, the consumer, to take that action. Below is a brief list of the ways your video can elicit a response from the consumer, can get a response from the viewer immediately, converting a passive pair of eyes into an active consumer.

Social Media Links
These are standard links to the most commons Social sites; Twitter, Facebook, etc. and usually have a variety of levels of direct connection options. These are usually CTA targets for Lead Generation, Learn More, or Social Sharing purposes.

Internal Commentary

Generally self-promoting CTAs, hoping to drive more Video viewing, comments on the video itself. Can also be great for testimonials, feedback and the like.

Embedded Links

Stronger means of linking in loyal users and consumers, so they help promote, expand viewership, and garner even more support from others.

End Credits Links

Most common for advertisers or vested interested in taking advantage of your growth, or for linking to drive more viewership. When you have enough drive, will have the opportunity for auto-starting the next of your videos.

Associated Website Annotation – for YouTube association with your site

Affords YouTube associates in good standing to put direct links and the like in the video frame itself. Good option if you are constantly driving more, and new traffic to your YouTube page.

Step Number Six: Get The Video Out There!

In this chapter, you will learn:

- How to get the video promoted before it gets released, to build excitement.
- Ways to expand your reach even after the video is out.

Build the Buzz

Email

When you decide to start a video functionality, start first with a pre-video email message to your proposed clientele, so they can be aware, and to prime the pump for their support.

Social Media

Moving on through the production phase, keep constant contact with your clientele through social media, releasing strategically snapshots and imagery of the happenings, to start the buzz in social circles, and get people talking.

Tagging

Along the way, release news stories and more images, tagging the photos with your sponsors, supporters, and the cast, in order to continue to get your clientele excited about the upcoming video.

Tweeting

The short-burst nature of Twitter is perfect for the daily updates and notes on the dailies, as well as for off-camera hijinks to keep the audience primed for the actual launch. Also great for the 'out takes and bloopers' aspect.

Embedding

Particularly for your loyal clients and supporters and sponsors, getting them early release embed links will get them to help you drive viewership.

Day of Launch Blowup Protocols

Now that the video is complete, laid out, linked, and prepared, the following steps are critical to get the buzz going right.

Website

○ LINK AND SITE DATA UPDATED FOR EASY ACCESS
Make sure your contact info, link buttons, CTA functions are all functioning. A blown link can botch the whole process. Be as perfect as possible, if you can. This is the moment of truth.

○ PRE-ANNOUNCEMENT ANNOUNCEMENT TO LOYAL CORE
You want every possible positive response you can garner, so do all you can to get the people who will support you to do so in your initial moment of glory.

○ SNEAK-PEEK LINKS TO THE INNER CIRCLE
Those of your inner circle, those closest to you, will do the most to expand the reach, so be sure to give them what they need to help you achieve your goals.

Social Media

○ LAUNCH ANNOUNCEMENT THROUGH ALL CHANNELS, ALL AT ONCE
Nobody likes mail bombs, but if you are going to do it, do it only once, across all spectrum of resource types. That way, even if it is uncomfortable to someone, it won't last long.

○ UPDATE INSTRUCTINS THROUGH A SINGLE CHANNEL
Once you have made the launch, choose one resource for all subsequent updates, to cut down on the spam, and to give everyone a place to go to chat about it, promote it, hear about it, etc.

○ LAUNCH UPDATE CYCLE IN PLACE.
The process of updating can even be automated, but be sure to meet your expected update times. Being late can be seen as unprofessional, or that you are holding data back. Be honest about your release times, numbers, etc.

Additional Information Release

○ PRESS RELEASE TO MEDIA OUTLETS
Before the actual release, let all appropriate news resources have access to the same numbers, the same information, all in a tidy promotional package. In particular, create the data in a palatable form. Many press release formats are available online, so choose one that matches your company's style.

o PRESS RELEASE WITH COVER LETTER TO INDUSTRY RESOURCES

When your business is part of a larger industry, you can do well to alert all the various tiers and channels into which your business fits.

o PRESS RELEASE TO LOCAL NEWS SOURCES

"Local Business Makes Good", the idea that your community will support your efforts because you are part of that community can certainly assist you in reaching people. Further, you may be able to generate local news stories and buzz by doing so, further expanding the success of your project.

Step Number Seven: Create Contemporary Content

In this chapter, you will learn:

- The broad spectrum of video possibilities, so you can choose the best for your message and your capabilities to create them
- Match your choice to your message for maximum impact.

What is happening NOW impacts the receptiveness of your video. Taking care to consider what kind of message you want to convey, what platform best matches it, and how best to go about its creation, you can hit the mark in a major way, and give yourself the greatest opportunity for success.

Determine the basis of your video

Since the video – creation field is so open, one would think that you could literally create anything as a video, and have a chance of becoming 'viral' with it. While there are precedents of this nature, where the person who put up the video did not think, nor even consider their video would become so appealing to so many, the vast array of successful videos fall into one of the following categories. As we look at each one, we'll recommend some categories for each that would be a good match, and some that probably would be better suited doing something else.

Prank or Practical Joke

o BASIC PREMISE:

Some setup is made and described to the viewer, with the idea that what you are recording are the witless public response to the action.

o VISUAL SATISFACTION:

Viewers enjoy wacky, real responses from the people captured on video, and the aftermath of discovery of the prank or practical joke

o RECOMMENDATIONS
Particularly good for discussion of human perception and good-natured response to people. Not particularly good for reaching out to possible clients who may equate the set-up as potentially reflective of your corporate honesty or candid nature.

Record a Monologue
o BASIC PREMISE:
Discussing or outlining a particular subject in a clear, humorous and direct way. Can be made entertaining by voice, inflection, costumes, conditions, etc. Offers an opportunity to say plainly what needs to be said without undue or irrelevant fanfare. Focuses viewers on the subject at hand.

o VISUAL SATISFACTION
Material is not cluttered by visual imagery. Can be humorous by intent or by accident, but generally is not particularly slapstick or goofy. Usually reflects positively on the company or individuals responsible for it.

o RECOMMENDATIONS
Perfect for new companies, with a serious vein. Make these shorter, rather than longer, or even serialize them for a stronger impact. Adding levity or some interesting things going on in the background, like Easter eggs – snippets or bits that the viewer will respond to (like some unrelated drama or such that will cause a reaction) which in turn will lead to more viewers looking for the egg, as it were. Be careful to not let that overshadow the message, though.

Lip-Sync Video
o BASIC PREMISE
Take a popular song or lyrical poem, and using voice-over, let others lip-sync.

o VISUAL SATISFACTION
Juxtaposition and irony go well to drive this popular type of video, regardless of the musical theme, and gets people thinking in new ways about what music and the lyrics mean.

o RECOMMENDATIONS
Particularly effective with young folks, and fans of whichever musical form you are using. Not effective or missing the mark if

dealing with material to which the lyrics or music are not related, but can be a great icebreaker video otherwise.

Adorable Family or Pet video

o BASIC PREMISE

Human nature sets us up for a visceral response to cute animals and endearing family situations.

o VISUAL SATISFACTION

One question: Who doesn't like cute animals, in our deepest heart of hearts?

o RECOMMENDATIONS

If the cutesy factor isn't too ridiculous, this will work for just about every category. The cleverer, the sweetest of these can go through the roof. Probably not a good match for a weapons manufacturer or mercenary force, but otherwise, this is good to go for most others.

Meme-Worthy Video

o BASIC PREMISE

A word, phrase, or situation that is eminently memorable, and that can be bent to your particular message or theme.

o VISUAL SATISFACTION

Links your message and concept to something that is already memorable, thereby driving your message deeper than it otherwise may penetrate.

o RECOMMENDATIONS

So long as the parody or meme themes are in line with what you want to convey, this can be very powerful. On the other hand, if there is little to associate them to one another, your meme concept may actually backfire, leaving a negative image. Use with caution.

Step Eight: Create. Promote. Release. Repeat.

In this chapter, you will learn:

- The deepest secret of video creation and production

- The Video release is a calculated and planned marketing program, as is any other type of promotion. It has a high and nearly unattainable topside, and a minimal expense and almost no downside.

Time Your Release

The one resource you can manipulate is the release date. Create the concept far enough in advance to be able to take into account all the variables.

Current Affairs

Keep your eye on what is happening around you, what news stories are around, what ways your video can reflect the topics and themes that would make it appealing to others.

Inclusion and Involvement

Reach out to community and business leaders involved, to get intrinsic support ready to roll when you are set for launch.

Seed the Release

Even before the actual release, keep the potential audience aware with teasers, outtakes, and other fun stuff, to drive interest in the video once available.

Coordinate for Instantaneous Review

One of the greatest tools in promotion is controversy. Have a couple of reviewers get a look in advance, and find one or more topics for them to talk about in a debate that does not detract from the message, but which engenders interaction.

Accelerate first 24 hours

If possible, dump new viewer markets over time, feeding off the excitement of one audience for the next, and the next. If your growth can reach critical mass within 24 hours, your ongoing success becomes far more likely.

Give incentives for comments and likes

We all want to support our friends, so do what you can to give people reasons to share or comment or like your video.

Bring a Friend, share a friend

Not only is it important to get folks to see and appreciate your video; it is critical that they connect and pursue your Call to Action. Get feedback from those that have been most positively impacted by your message, and get that feedback into the loop ASAP. People do what they see others do successfully.

Be prepared to Fuel the Fire

Even after your video has been released, there is much you can do to drive new viewers.

Bloggers

- CONTACT CONSUMER-DRIVEN SITES

  While it takes the longest to drive one way or another, starting early and repeating often your message to the public is critical. More eyeballs means more traffic, more CTA response, more of everything you need for the campaign to be successful

- CONTACT INDUSTRY ALLIES

  Reaching out to sympathetic and similar sources is at once an acknowledgment of the industry as a whole, but a bit of good natured quid pro quo will help drive additional interest.

- GENERAL INTEREST AND MASS MEDIA

  While generally you want to stay focused on your target demographic, it is also good to create a general awareness of your campaign, your message, your product or service.

Video Review

- SPECIALTY REVIEW

  Your marketplace has its own experts. Work with those individuals and teams with a good reputation, to get a positive or at least affirming message from them that you can use to further promote and expand attention to your campaign.

- CATEGORICAL REVIEW.

  From a production standpoint, getting identity supported by others in the same kind of project can further build rapport with others, and perhaps build an ancillary viewership through the association therewith.

○ GENERAL REVIEW
Much as dealing with consumers directly, choosing to work with promotional Websites and those that are actively seeking new and innovative video gives you a lift, just for attempting the process.

Conclusion: Do's and Don'ts

Do's

Specialize in Content Theme or Message
The better you become at what you do, the greater the chance others will see the value in what you are accomplishing

Use quirky humor and irony
With your tongue firmly in cheek, you have a better chance of weathering the current field of videos being created. Humor and positivity are the objectives now, and you will do well to take advantage of this trend.

Develop Your Strategy beyond Your Boundaries
Maybe your humor is too biting for your current clientele. Realize that there are a wide array of tastes, and thinking beyond your current framework will offer opportunities you might otherwise miss.

Perform Due Diligence on material
Clear the IP barriers by remaining free of encumbrances of plagiarism, libel, etc. Make sure your legal standing is iron-clad.

Open your boundaries and restrictions as much as possible
Inclusiveness is a huge benefit in the video marketplace. Consider all the alternatives, and see how you can take advantage of that openness

Keep producing
It has taken this far to actually say it, but your best likelihood on creating a successful video is not instantaneous viral acceptability, but of making a long series of successful videos, that do what you desire of them, that convey your proper message, and that get the appropriate response to your calls to action. DO good work, and you will achieve the appropriate measure of success.

Don'ts

Become subject to your own press
It is pretty easy to watch growth happen, and believe your job is done. But a constantly repetitive video business is more than that. Stay controlled about our business, keep your outreach expanding. Keep a rational eye on your work, and don't get fooled by instantaneous success or failure. All things change.

Take criticism personally
A reviewer's venom is often just a manifestation of their emotional state, and rarely is reflective on the material they review. Just as you don't want to overblow your praise, take into consideration the origins of the complaint or criticism, and look only for warranted improvement. In any negative there is a root of truth, so take that into account on future projects.

Be above the parody
If a video you do gets spoofed, take it as high praise, no matter its take on your version. The benefit of replication stems from the reality that such will lead inevitably back to the source, thereby driving your video along with theirs. Even consider making spoof videos of your own video, since you can see the flaws that the parody inevitably draws from.

Forget to acknowledge others
The worst thing one could do is take sole credit for something that is clearly a collaboration. Getting the word out about your awesome team, all the help you have received, and who all was involved are all ways to improve your standing, and drive new loyal viewers to your videos.

Forget to monetize
In all of your video production, remember what it is you are trying to accomplish. When possible use Google AdSense and AdWords programs to drive revenue from viewership, and make the best out of your promotional needs.

Stop making more
Don't see an individual video as being your sole objective. An ongoing campaign of new videos, expanded viewership, and a

broader consumer base for your products and services needs to be your primary concern.

Specific Details & Common Terms in Video Production

Action Axis
The imaginary line along which subjects move to maintain screen direction. Crossing it or reversing it causes errors in continuity that are confusing to the viewer.

Close-up (CU)
Tightly-frames camera shot where the principle subject is viewed at close range. Pulled back would be considered "medium close up" – (MCU), while zoomed in very close is "extreme close up" – (ECU or XCU)

Composition
Visual constitution of all characteristics of a shot. Combined qualities form an image that is pleasing to view.

Depth of Field
Range in front of a particular camera's lens in which images appear in focus. Depends on distance from subject, focal length of the lens and aperture settings.

Establishing Shot
Opening image of a program or scene. Usually wide angle or distant perspective that orients the viewer to the setting and surroundings.

Framing
Process of composing a shot for content, angle, and field of view as well as perspective.

Long Shot (LS)
Camera view of subject or scene from a distance, showing broad perspective

Nose Room
Space between subject and the edge of the frame, in the direction the subject is looking. Also known as 'look room'.

Over-the-shoulder shot
View of the subject with another person's shoulder in the shot. Used in interview shots and situations.

Pan
Horizontal camera movement, from Right to Left or Left to Right, from a stationary position

Pedestal
Vertical camera movement, from low to high, or high to low, keeping the cameral level.

Rack Focus
Shifting focus between subjects in the foreground and background so the viewer focus shifts with the camera.

Remote
Video shoot performed on location, outside a studio environment.

Rule of Thirds
Composition theory wherein field is divided horizontally and vertically by thirds, and subjects are placed along those lines.

Scene
In the language of moving images, a sequence of related shots usually consisting of all action in a particular location.

Shot
Intentional, isolated camera view that will work with others to create a scene.

Tilt
Vertical rotation as if on a tripod.

Tracking
Lateral movement of a camcorder that travels with a subject, Should remain a uniform distance from the subject.

Vignette
Visual special effect where viewers see them image through a perceived keyhole, heart shape, etc. Low-budget films can attain this effect through aiming the camera through the appropriately-shaped cutout.

Whip Pan
Extremely rapid camera motion from left to right, or vice versa, appearing as an image blur. Two such pans in the same direction edited together to make one moving from, another moving to a

stationary shot can effectively convey a passage of time, or a change of location.

With this working vocabulary, these tips, and a recommendation to just go make some videos, you are on your way. Who knows, you may have the next Grumpy Cat! And even if you don't ever succeed in making the Viral Video, you will at least have quality video in your library, and a host of new skills to deploy on your next big project!

So share this information, and get behind the camera. There is a world to record out there.

To hear about Entrepreneur Publishing's new books first (and to be notified when there are free promotions), sign up to their New Release Mailing List.

Finally, if you enjoyed this book, please take the time to share your thoughts and post a review on Amazon. It'd be greatly appreciated!

Thank   you   and   good   luck!

Preview Of 'Kindle Publishing For Entrepreneurs: 9 Steps To Producing Best Selling Amazon Kindle Books And Building Incredible Passive Income' from Entrepreneur Publishing

Defining Your Audience

It is very important to be clear about your target audience, i.e: the individual or group you wish to sell your product to. The book you write must then be written keeping your target audience in mind so that they are interested in buying it and can also enjoy its benefits.

When writing an eBook you should have some ideas on how to attract your audience. Online readers are not very patient people when it comes to looking for something and they only stop to read something that looks attractive to them, and it is evident that you can't have sales if no one reads your eBook. To make your job interesting and to attract the right target audience is not that hard of a task, you just need to know the right buttons to push. You should be able to define your target audience by the points explained below.

Demography

This is focused on the socioeconomic characteristics of people in a given region. When it comes to profiling the audience that you plan on offering your products to, you should ensure that they are interested in what you have to offer and they would be willing to pay for it. Demographic information will guide you well in this endeavor and can even give you ideas of another business after studying the people in a certain region and realizing that they can consume some other product too. Here are some demographic factors that you should put into consideration while looking for an audience:

- Age

You must be well aware of your target audience's age. For example, if you are targeting children then you will have to use simple vocabulary and avoid erotic stuff. On the other hand, if you are writing for teens then you should have things that they are interested in such as romance and fantasy.

- Gender
There are things that women love to read about that men don't and vice versa. So when you are looking for the right target audience, first decide which gender you wish to target. However, there are some topics that are well in demand by both men and women.

- Occupation
You can tell a lot about someone by the type of job they do. If you have some high profile job, chances are, you're a sophisticated person, and hence you will be interested in specific topics. There are several eBook ideas based about certain occupations, such as tips etc. hence this point is of importance.

- Education
Literate and illiterate people have a slightly different taste in many things, for example when it comes to reading, the illiterate cannot buy books which they cannot read. For you to be selling an eBook means that the audience that you target should be literate.

- Marital Status
This is very important because your topic selection largely depends on your target audience's marital status. For example, if they are married you can write content related to husband-wife issues, bringing up a child etc.

- Income Level
This will give you a very good idea on who to sell to, what to sell and at what price. People who earn a lot of money tend to spend on lavish items and can afford to pay more. Finding a target market of high income people and finding some of the

commodities that they are much interested in will see you make some decent sales.

- Average Family Size
  If you are looking to be selling eBooks and your target audience is made up of huge families, it would make sense to sell books that are relevant to kids so that the parents can purchase some for their kids. With the right products, large families can be really great markets for one to invest in.

All this information will help you come up with custom content that is fit for the audience that you just profiled. At times you might find that within one big geographical area, there is more than one profile of people and this can be a great idea for you to invest in. But understand that demographics are just one part of the solution, you need to pay attention to several other things.

Address The Consumers Questions

A consumer who goes online to look for products that they want to buy often gets confused and ends up having numerous questions. There are a great number of online retailers and hence the buyer often gets confused about the selection of a store. It is your work as a retailer to ensure that you address these questions and give comprehensive answers so that the consumer can trust and want to buy from you by picking you over other alternatives.

When dealing with people especially on social network platforms you should use content of high quality. Answer their questions and then help the consumer solve whatever issue they might be facing. Good and informative solutions will lead to them sharing your solutions to their friends and this will create a much bigger fan base for you and that results in more potential clients. Writing and publishing eBooks that help common people solve their day to day issues could be a huge hit with a large customer base.

Understand The Potential Consumer

When profiling a group of people that you have the hopes of making your clients, you should first understand them better. This has been made easier today by the use of social media networks; you can learn a lot about someone even without talking to them. You can learn someone's favorite things, places that they like to visit social media platform that they like using and the ways that they prefer when gathering information. Look to understand what the customer cares about for; if you are able to show that you understand and share their concern, you will be able to create a long-term relationship that will be beneficial to the both of you.

Click here to check out the rest of Kindle Publishing For Entrepreneurs: 9 Steps To Producing Best Selling Amazon Kindle Books And Building Incredible Passive Income on Amazon.

Or go to: http://amzn.to/1a2iokG

More Books for Entrepreneurs

Click here to check out the rest of Entrepreneur Publishing's books on Amazon.

Below you'll find some of my other popular books that are popular on Amazon and Kindle as well. Simply click on the links below to check them out. Alternatively, you can visit my author page on Amazon to see other work done by me.

How Audiobooks Make You Smarter: 7 Little Known Ways Audio Books Can Boost Memory Capacity And Increase Intelligence

How To Write A Book And Publish On Amazon: Make Money With Amazon Kindle, CreateSpace And Audiobooks

Gardening For Entrepreneurs: Gardening Techniques For High Yield, High Profit Crops

Speed Reading For Entrepreneurs: Seven Speed Reading Tactics To Read Faster, Improve Memory And Increase Profits

Content Marketing Strategies: How Delivering Sensational Value Can Help You Build A Digital Media Empire

Kindle Publishing For Entrepreneurs: 9 Steps To Producing Best Selling Amazon Kindle Books And Building Incredible Passive Income

If the links do not work, for whatever reason, you can simply search for these titles on the Amazon website to find them.

www.ingramcontent.com/pod-product-compliance
Lightning Source LLC
Chambersburg PA
CBHW071016180526
45168CB00003B/1448